NATIONAL
GEOGRAPHIC
KiDS

GO WiLD!

Pandas

Margie Markarian

NATIONAL GEOGRAPHIC
WASHINGTON, D.C.

What's black and white
and fluffy all over?

If you guessed the giant panda bear,
you are right!

Pandas are cute, curious, and cuddly.
Let's visit the world of pandas together.

Up! Up! Up!

Pandas live in bamboo forests, high in the mountains of China.

Their habitat has tall trees to climb, rocky ledges to explore, bamboo plants to eat, and grassy spots to take a tumble.

WHEE!

Panda Land

China is the fourth largest country in the world. It is on the continent of Asia.

Pandas live in the middle of China, both in the wild and on nature reserves such as the Wolong National Nature Reserve.

Pandas also live at zoos in several countries around the world.

CHAPULTEPEC ZOO, MEXICO

SMITHSONIAN'S NATIONAL ZOO, U.S.A

BEAUVAL ZOO, FRANCE

UENO ZOO, JAPAN

CALGARY ZOO, CANADA

Size Matters

Pandas are called "giant pandas" for a good reason— they're BIG!

A full-grown panda bear weighs 200 to 300 pounds (91–136 kg). From head to back paw, it measures four to six feet (1.2–1.8 m) long. That's about the length of a couch.

Not Just Any Body

Pandas are adorable—just look at that round, furry white face, those dark eye patches, and those fuzzy black ears! From head to toe, the panda's body is equipped to handle life in misty mountains and deep forests.

Black fur helps pandas hide in the forest. White fur helps them hide in the snow.

Thick, woolly fur keeps pandas dry and warm in cold, wet weather.

Strong jaws and large, flat teeth are good for crushing and chewing tough bamboo.

Sharp claws are handy for grabbing bamboo stalks and for hanging from branches.

Padded paws prevent slipping on wet surfaces.

13

Thumbs-Up!

A panda's paw has an extra-long bone that sticks out like a thumb. This special body part makes it easy for pandas to GRAB and GRIP bamboo.

Meet the Cousins

In addition to pandas, there are seven other kinds of bears.

Brown Bear
Most brown bears in North America are called grizzly bears.

American Black Bear
These bears are not just black! They can also be brown, beige, blue-gray, or white.

Sun Bear
The sun bear's long tongue is perfect for licking up lots of termites and bees. Yum!

Spectacled Bear
Light fur around this bear's eyes looks like spectacles, a fancy word for glasses.

Sloth Bears
Don't let the word "sloth" in this bear's name fool you. Sloths are famous for moving slowly, but sloth bears run fast!

Polar Bears
Thick white fur and a layer of fat keep polar bears warm in super-cold weather.

Asiatic Black Bear
This bear's nickname is moon bear, for the white crescent moon shape on its chest.

Bamboo Bites

When it comes to eating, pandas have a one-track mind: They eat bamboo for breakfast, lunch, and dinner.

Any type of bamboo will do.
Any part will do, too—stems,
leaves, shoots, or flowers.
It's all yummy to pandas!

CHOMP! CHOMP! CHOMP!

Search! Snack! Snore!

Pandas find bamboo delicious, but it's not that filling. They need to eat 25 to 40 pounds (11–18 kg) a day to keep their tummies full and their bodies healthy.

A panda spends about 14 hours a day in search-and-munch mode.

In between snacks, they take short
SNOOOOOOZZZZZZES.

Sweet dreams!

Hey, Baby!

Newborn pandas are called cubs. Cubs are TEENY TINY. They weigh less than one pound (0.5 kg) and measure five to seven inches (13–18 cm) long. That's about the size of a banana!

Like other mammals, newborn pandas drink their mother's milk. Panda moms stay close to CUDDLE and SNUGGLE.

At birth:
Newborns are pink, blind, and helpless.

3 weeks:
Peekaboo! Black-and-white fur starts to show.

6–8 weeks:
Eyes open. Hello, world!

3 months:
Crawling and scooting—what a cutie!

26

4 months:
Wobbly walking
on all fours.

5–6 months:
Ready to chew
on bamboo and
climb sky-high!

1 year:
50–60 pounds
(23–27 kg) of
roly-poly fun!

2 years:
Bye, Mom. I'll be
OK on my own!

Party of One

Giant pandas are solitary animals. That means they live alone.

But pandas still need to communicate with each other. Sometimes they communicate by leaving their smells on rocks and trees.

All it takes is a
SWISH, SWISH
of the tail to send a
smelly message!

Say What?

Pandas also communicate by making sounds. They don't ROAR like other bears, but they make lots of other noises.

SQUEAK! SQUEAL!

HONK! HUFF!

CHIRP! MOAN! GROWL!

BLEAT! BARK!

Giant panda moms are 900 times bigger than their newborn babies.

Panda moms carry around their cubs in their mouths.

The Chinese word for panda is *daxiongmao*. It means "large bear-cat." Here's how you say it: da-shee-ONG-ma-oh.

Pandas at zoos eat more than bamboo. They are fed carrots, apples, and sweet potatoes, too. Sometimes, ice pops made from frozen fruit are on the menu!

In keeping with a Chinese tradition, a panda cub born at a zoo or reserve is not named until it is 100 days old.

SPLASH! Pandas are excellent swimmers.

Panda poop has so much fiber in it that it can be made into paper.

Pandas live to be about 20 years old in the wild and about 35 years old at zoos.

Panda Problem

Long ago, there were lots of pandas.
But as cities grew, forests were cut down.
Many pandas lost their homes.

Today, there are only about 1,800 pandas left in the wild. Habitat loss is still a problem for pandas.

To the Rescue!

People are doing many things to help pandas, such as creating safe places for them to live. The Wolong National Nature Reserve in China is the biggest panda sanctuary. It has a research center where people study pandas and help panda moms give birth to strong, healthy cubs.

Are You My Mother?

Cubs who get used to seeing people have trouble living on their own in the forest. So caregivers at reserves sometimes wear sticky, stinky panda costumes.

The costumes are scented with panda poo
and pee so they smell like panda moms.

Protect Pandas!
You can help pandas, too.

Start by sharing what you know and love about pandas with friends. This is called **raising awareness**.

You can also talk to an adult about collecting coins for *Pennies 4 Pandas*. An organization called Pandas International uses the money to save pandas' habitat.

Every penny counts!

Name That Animal!

Giant pandas just might be the cutest black-and-white animals, but they aren't the only ones!

Can you name these other black-and-white animals?

1

2

Want to build your child's enthusiasm for pandas?

Visiting pandas at a zoo is a great way to start. Pandas live at zoos in three cities in the United States—Washington, D.C.; Atlanta, Georgia; and Memphis, Tennessee—as well as in numerous countries around the globe. Even if there is not a zoo with pandas near you or anyplace you plan to travel in the near future, you and your child can still go on a virtual visit. Many zoos and reserves have panda cams that can be viewed 24/7. And be on the lookout for opportunities to participate in Name That Panda contests. Zoos often celebrate the birth of new pandas by inviting people to give suggestions. Here are some other activities for you and your child to do together.

Panda Pretend Play (Craft and Movement)

Your child can pretend to be a panda with a do-it-yourself costume. To make a panda mask, all you need is a sturdy white paper plate, black construction paper, a black crayon, and two pieces of yarn. Help your child cut two holes for the eyes and then draw black patches, a nose, and a mouth. Cut black circles for round ears and glue each one to the top of the plate. Punch a hole on each side of the plate, pull a piece of yarn through each, and knot. Complete the look with a white shirt and black pants or leggings. Now you're ready to roll, tumble, and somersault!

Show-and-Tell!
(Public Speaking and Writing)

For a show-and-tell day at school or a family gathering, suggest your child bring this book and a picture of a panda and/or a stuffed animal, and share five panda facts. Take pictures of the performance and have your child create a small scrapbook with short sentences to go with the visual highlights.

Adopt a Panda
(Responsibility)

No, your family doesn't actually get to bring a panda home, but adoption programs give you and your child a way to help a panda have a safe and healthy life. Both Pandas International (pandasinternational.org) and the World Wildlife Fund (wwf.org) have "adoption" programs based on different levels of giving.

Bamboo Weigh-In
(Math)

What does 25 to 40 pounds (11–18 kg) of bamboo "feel" like? Use a shopping trip to put it into perspective for your child by pointing out various items and what they weigh. Start with a one-pound (0.5-kg) carton of butter and go up from there: a five-pound (2.2-kg) bag of flour; 10 pounds (4.5 kg) of potatoes; a 15-pound (6.8-kg) box of powdered detergent; two gallons of paint (about 20 pounds [9 kg]); six reams of printer paper with 500 sheets each (about 30 pounds [13.6 kg]); and a large bag of dog food (about 40 pounds [18.1 kg]).

Once Upon a Time
(Literature and Cultural Connections)

Reading about real pandas can spark an interest in fiction and folklore starring pandas. Try *Legend of the Giant Panda* by A. B. Curtiss, a charming picture book that relays a tale about how pandas got their black-and-white coloring, as well as a message on forest conservation. Little panda fans may also enjoy *The Pandas and Their Chopsticks*, a delightful folktale about generosity by Demi, and *Goldy Luck and the Three Pandas* by Natasha Yim, a panda-centric retelling of the classic story.

GLOSSARY

bamboo: a plant with a hollow, woody stem

climate: the usual weather in a place over a period of time

communicate: to pass on information

continent: one of the seven main landmasses on Earth

cub: a baby bear

fiber: material in a plant that cannot be broken down (digested) by the body

habitat: an animal's natural home

hibernate: to sleep for the winter

mammals: a group of animals, including humans, that have backbones, are warm-blooded, breathe air, have hair or fur, and drink milk from their mother

research center: a place where people study, discover, and report new ideas and information

reserve: a protected area of land or water

For Dad, who loved watching animal shows and sharing fun animal facts —M.M.

AL=Alamy Stock Photo; GI=Getty Images; MP=Minden Pictures; SS=Shutterstock

Front cover, Hupeng/Dreamstime; back cover, Mohd Rasfan/AFP/GI; (leaves throughout), Kolonko/SS; 1, Yoreh/Adobe Stock; 5, Mohd Rasfan/AFP/GI; 6 (LE), Gavin Maxwell/Nature Picture Library; 6 (RT), Hung Chung Chih/SS; 7 (LE), bgspix/SS; 7 (UP RT), Keren Su/China Span; 7 (LO RT), mauritius images GmbH/AL; 8, NG Maps; 9 (UP), Pedro Pardo/AFP/GI; 9 (CTR LE), Erika Bauer/AP/SS; 9 (CTR RT), Eric Baccega/MP; 9 (LO LE), Yoshikazu Tsuno/Reuters; 9 (LO RT), Dave Pattinson/AL; 10-11, Pascale Gueret/SS; 12, Isselee/Dreamstime; 13, Eric Isselee/SS; 14, Katherine Feng/MP; 15, clkraus/SS; 16, Henrik Karlsson/NiS/MP; 17 (UP), Jeanninebryan/Dreamstime; 17 (LO), wrangel/iStock/GI; 18 (UP), Juan Carlos Vindas/GI; 18 (LO), Christian Hutter/AL; 19 (UP), Pär Edlund/Dreamstime; 19 (LO), Engdao Wichitpunya/AL; 20, Jean-Paul Ferrero/Auscape/MP; 21, Bryan Faust/SS; 22, Sylvain Cordier/Biosphoto; 23, Glow Images/GI; 24, Katherine Feng/MP; 25, Mitsuaki Iwago/MP; 26 (UP), Lola Levan/EPA/SS; 26 (LE), WENN Rights Ltd/AL; 26 (LO), Xinhua/AL; 26 (RT), Li Qiaoqiao/Xinhua/Alamy Live News; 27 (UP LE), Katherine Feng/MP; 27 (UP RT), Katherine Feng/MP; 27 (LO LE), Katherine Feng/MP; 27 (LO RT), Mohd Rasfan/AFP/GI; 28-29, Jenny E. Ross; 30, Pascale Gueret/iStock/GI; 31 (UP), Pascale Gueret/iStock/GI; 31 (CTR RT), Mitsuaki Iwago/MP; 31 (LO), Suzi Eszterhas/MP; 32-33, Pete Oxford/MP; 34, View Stock/AL; 35, Joseph Van Os/GI; 36, Suzi Eszterhas/MP; 37 (UP LE), Katherine Feng/MP; 37 (UP RT), Yong Wang/SS; 37 (LO LE), Pongmanat Tasiri/EPA/SS; 37 (LO RT), Eugene Hoshiko/AP/SS; 38, China Photos/GI; 39, China Photos/GI; 40, Eric Isselee/SS; 41, Nagy-Bagoly Arpad/SS; 42 (LE), Clara Bastian/SS; 42 (RT), Volodymyr Burdiak/SS; 43 (UP LE), Robert Tripodi/SS; 43 (UP RT), Galina Savina/SS; 43 (LO LE), JeremyRichards/SS; 43 (LO RT), ArCaLu/SS

Designed by Kathryn Robbins

Hardcover ISBN: 978-1-4263-7160-8
Reinforced library binding ISBN: 978-1-4263-7161-5

The publisher would like to thank Dr. David Kersey for lending his expertise on this animal. The publisher would also like to thank Angela Modany, associate editor; Sarah J. Mock, senior photo editor; Mike McNey, map production; Liz Seramur, photo editor; Anne LeongSon and Gus Tello, design production assistants; and Alix Inchausti, production editor.

Printed in Hong Kong
21/PPHK/1